T0023204

THE STORY OF
HARLEY-DAVIDSON

Published in 2022 by Welbeck
An imprint of Welbeck Non-Fiction Limited,
Part of Welbeck Publishing Group.
Based in London and Sydney.

www.welbeckpublishing.com

A CIP catalogue record for this book is available from the British Library.

ISBN 978 1 80279 294 2

Editors: Ross Hamilton & Conor Kilgallon
Design: Russell Knowles & Luana Gobbo
Picture Research: Paul Langan
Production: Rachel Burgess

Printed in China

10 9 8 7 6 5 4 3 2 1

THE STORY OF
HARLEY-DAVIDSON

A TRIBUTE TO AN
AMERICAN LEGEND

JOHN WESTLAKE

WELBECK

CONTENTS

EARLY YEARS

STARTING WITH SINGLES

The beginning of the 20th century was an extraordinary period for motorcycling, with would-be engineers working on all manner of two-wheeled devices in sheds and workshops around the world. Hundreds of new companies sprang up, each trying to tempt customers away from their horse with affordable, speedy, reliable transport. Most withered almost immediately, but in Milwaukee, USA, two young men were destined to beat the odds in an almost unimaginably successful fashion.

William S Harley and Arthur Davidson had been friends since school and were both fascinated by engineering. After leaving school, Harley worked as a draughtsman for the bicycle company where Davidson was an artisan, creating original parts that would then be cast or forged for production. In terms of their skills, the pair were well suited to starting an engineering empire.

In 1901, the 20-year-old Harley drew up plans for a 106cc 'bicycle motor'. Records of these early days are sketchy, but it's thought the design started life as a kit engine based on a French De Dion-Bouton motor (the French were an engineering powerhouse back then).

OPPOSITE: A Silent Gray Fellow, this one a single-cylinder machine from 1912.

ABOVE: It's 1907 and William S Harley (third from left) and Arthur Davidson (far left) have been joined by two more Davidsons: Walter (second left) and William (far right).

More certain is that the two lads were assisted by Davidson's brother Walter, a skilled mechanic who worked as a railroad machinist. He added the impetus needed to finish the first prototype, which was completed in 1903. Although no photographs of the machine exist, we know it wasn't much to get excited about – little more than a bicycle with William Harley's small engine attached. Inevitably, the performance was poor, and it wasn't even able to climb Milwaukee's modest hills.

MORE POWER!

Despite the inauspicious start, Walter was enthused enough to join the team, and another machine soon followed – with a far bigger engine. Looking at the design of this 405cc single cylinder, it seems likely our heroes worked closely with Ole Evinrude, another young Milwaukee engineer who went on to create the vast Evinrude outboard motor company. There are several similarities between the Harley-Davidson (the company was officially incorporated in 1903) and Evinrude motors of the era and some commentators believe he was responsible for the Harley engine's roller tappets.

The team were also influenced by the new Merkel bikes that appeared in late 1902, adopting the Merkel's frame design which looped down and round the engine rather than being a straightforward bicycle-style diamond. Harley's second model appeared in 1904 and is regarded by many (including the official company historians) as the first real Harley-Davidson motorcycle. The 1903 bike was merely a 'powered bicycle'.

And so the adventure began: a Chicago motorcycle enthusiast liked the second model so much he offered to sell

BELOW: This Serial #1 was not the first Harley – there were prototypes before – but it is definitely the oldest one in existence. It curr ently sits in the lobby of Harley's Milwaukee HQ.

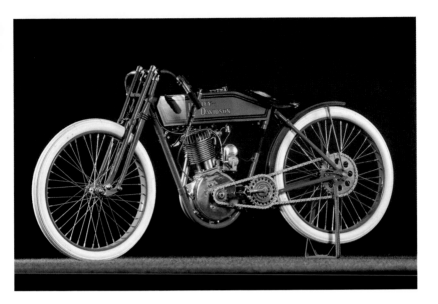

them in his home town and bought three in 1905, then 24 the following year. In 1906 the team built a factory on Juneau Avenue, where Harley still has offices today, and in 1907 Arthur's older brother Bill Davidson joined to run the machine shop. The clan was growing.

Ever since the first motorcycles appeared, owners had raced them, and as the industry grew, so did the importance of sporting success. The adage 'win on Sunday, sell on Monday' was coined in the 1960s but could just as well have applied to Harley's early years. Walter Davidson had won a few trophies in regional races, but when he scored a perfect 1,000 in the National Endurance Contest held over 356 miles in New York State, the American public took notice. Harley's advertisements of the day played on the reliability and rugged build quality needed for such a win, but oddly it was 1914 before Bill Harley started a factory race department.

ABOVE: Harley's single-cylinder board track racers were visions of simplicity. This 9B is from 1913.

OPPOSITE: In 1915 Harley's adverts sold the respectability and reliability of the company's first V-twin, the Model 7D. Like other early Harleys it was advertised as a Silent Gray Fellow because of its effective exhaust muffler.

ENTER THE V-TWIN

Though the modern-day Harley-Davidson is famous for the V-twin, in the early days the company was sluggish to get behind the engine layout – despite it powering Walter to his National Endurance win. Their single-cylinder bikes were selling so well that although Harley did advertise its 880cc Model D V-twin in 1907, not many were actually built because the intake system on pre-1911 bikes made the bike unreliable. The twin was effectively two of the single-cylinder motors joined together, and the single's suction intake valves didn't work well pulling air through one carburettor.

Once that issue was solved by using a pushrod and rocker to activate the intake valves, the V-twin took off. Soon the bikes came with newfangled rider aids such as lights, a clutch, kickstarter and gears. Before that all Harleys had one gear, no clutch, and were started by running along and jumping on board. To stop, you had to stall the engine.

By 1915, the Model F – a 989cc V-twin with three gears – outsold all Harley's other models combined and by 1916 Harley's total sales were 16,924: a staggering number for a company that was barely over a decade old.

The factory itself had changed beyond all recognition. From little more than a double garage-sized workshop with three men hand-building prototypes, Harley-Davidson's factory had become a 28,000 sq m (300,000 sq ft) operation employing 1,500 people in just ten years. However, despite these numbers, it's worth remembering that Harley wasn't yet the dominant force it is now – in 1913 the company produced just 18 per cent of American-built motorcycles. The war was about to change all that.

ABOVE: By 1921, Harley's Model JD was not just fast, but practical too – note the lights, mudguards, rack and girder front suspension.

WORLD WAR I STRIKES

Though America entered World War I in 1917, Harley-Davidson wasn't approached by the US military to supply bikes until 1918, when it made 8,000 for the army. Harley went on to supply at least 12,000 more before the end of the war in 1918. The bikes performed well – helping cement Harley's reputation for reliability – and when the war finished, the company found itself as the world's largest motorcycle manufacturer.

This remarkable turn of events was partly due to the European manufacturers having been tied up with war work for longer than Harley, partly thanks to a $3 million bank loan allowing a huge investment in production capability, and partly because of American rivals failing to make it through the war. But it also had a lot to do with Harley's reputation for rugged reliability going global – after the war nearly 20 per cent of Harley's production was earmarked for export.

Riding a post-war high, Harley launched the 1213cc (74 cubic inch) Big Twin engine in 1921, fitting it into new JD and FD models. These 18bhp bikes were a hit, but storm clouds were brewing.

Cars were stealing sales from bikes, not helped by Henry Ford cutting the price of his Model T to the same as Harley's biggest twin. Given that almost every buyer bought their motorcycle as transport rather than a pastime, this was a serious blow. Plus, a recession was taking hold. Harley posted a loss for the first time in 1921 – selling just 10,202 bikes, down from 28,000 the year before. It would be 21 years before sales reached 1920 levels again. Mind you, it could have been worse. At one point there were over 100 American motorcycle manufacturers in business, but the recession did for most and by 1928 only Harley, Indian, Henderson, Cleveland and Super-X (formerly Excelsior) were left.

BELOW: In the 1920s, board track racers like this FD were the fastest, lightest and most extreme motorcycles out there. And the FD was quick – in 1921, one of Harley's riders averaged over 100 mph during a board track race.

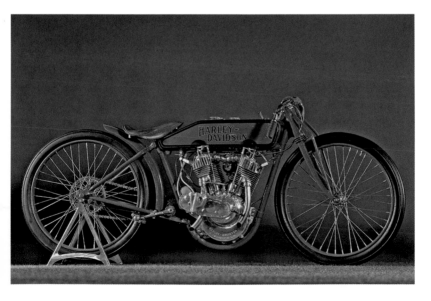

Harley's founders struggled to decide which way to go. Should the company concentrate on police and courier sales, or try to compete with an influx of high-tech high-revving British motorcycles? In the end they attempted to do both by building a 350cc single-cylinder model in 1926 to complement the Big Twins. It came in two forms – an economical side-valve and a high-performance overhead valve version – but both were unsuccessful. American buyers didn't seem to want little bikes, and new tariffs ruined their chances abroad where smaller bikes were more popular.

But Harley kept innovating, with both their engineering and product line-up. Having entire model ranges stemming from one engine is commonplace now, but in the 1920s it was rare, and Harley led the way – in 1928 the company had three basic engines yet created 12 separate models from them. And Harley's development engineers were busy too, introducing the first ever front brake on a production motorcycle in the same year.

RISE OF THE FLATHEAD

From the early days right up to 1929, Harley used an F-head engine design. This has the inlet valve on top of the combustion chamber, with the exhaust valve in a side channel, and though rudimentary compared to modern engines it worked well. However, the engines were expensive to make because of the complex combustion chamber, rocker arms and pushrods, plus they were difficult to keep reliable and a pain to service.

In 1929 they launched the DL, a 742cc (45 cubic inch) side-valve V-twin, which as the name suggests had all the valves to the side of the cylinder. This cured a lot of the ills of the F-head and meant the top of the cylinder could be kept slim, hence the name: Flathead.

The DL was intended as a halfway house between the big twins and the more agile singles. Though lacking the power

OPPOSITE: Harley's Flathead first arrived in 1929 and was a step up from the old F-head thanks to increased reliability.

and glamour of the bigger twins, the DL hit the spot with
customers who couldn't afford the bigger capacity machines.
Harley sold 2,343 DLs in 1929 at $290 each and the bike was
crucial in keeping the company afloat over the coming years.

In 1930 Harley unveiled a 1200cc (74 cubic inch) version
of the Flathead engine, known as the 74. This produced 28bhp
– 15 per cent more than the old F-Heads – and it was installed
in the new VL model. Like the DL, the easy to build and
maintain VL sold well despite the engine's extra weight taking
the edge off acceleration.

One reason for its success was that once some significant
teething troubles were overcome, the VL was a lot more reliable
than previous models. This meant riders could take advantage
of the American government's vast investment in the highway
network and travel further and faster with fewer fears of being
stranded by the roadside. Long-distance touring became
increasingly popular in the 1930s, and the Flatheads were
perfect for the job.

THE GREAT DEPRESSION

Then, just as things seemed to be picking up for Harley, along came two nightmares. First, a patent lawsuit went against it and the company had to hand over $1.1 million to a rival named Eclipse because a Harley clutch design was deemed to infringe a patent. But more serious was the Great Depression, triggered by the 1929 Wall Street Crash. It was economic carnage – sales plummeted, 1,300 banks closed and by 1930 Harley's factory was operating at just 10 per cent capacity.

Millions of people were made unemployed and unsurprisingly few of those in jobs had the confidence to splash out on a new bike. Harley's management team cancelled a V4 project and battened down the financial hatches: the mission was purely to survive. It was a wise move: rival companies Cleveland, Henderson and Super-X did not make it through this fraught period.

BELOW: The VL 1217cc side-valve model arrived in 1930 – just in time for the Great Depression. This 1933 model is a VLE Sport Solo.

DESPERATE TIMES...

In the wake of the Great Depression, the Harley team came up with some good ideas. They kicked off a scheme to sell branded clothing and accessories, little knowing that it would eventually become a multi-million-dollar enterprise on its own. The company also launched the Servi-Car, a three-wheeler powered by the 45 Flathead motor aimed at delivery companies and the police. Modern Harley riders might be surprised by such a tactic, but back then huge numbers of Harleys were utility vehicles, often with sidecars, so creating what was essentially a three-wheeled van was no great stretch. Harley still has three trikes in its range today, so it hasn't abandoned the idea of three-wheelers. Despite being rushed out to help bring

BELOW: Cheaper than a van or a truck, the Servi-Car proved popular with police forces as well as delivery companies. Some were still being used by the police in the 1980s.

in some much-needed cash, the Servi-Car was a solid design and, amazingly, stayed in the Harley range for 43 years, only being dropped in 1974.

KNUCKLEHEAD HERALDS A NEW ERA

In 1936 Harley made a huge advance with the introduction of the overhead valve 1000cc (61 cubic inch) Knucklehead. This was a momentous engine for Harley, one which would provide the template for half a century of success. And the Knucklehead was evidence that Harley had moved on from survival mode: it was determined to climb out of the Great Depression by building superb motorcycles.

ABOVE: The Flathead-powered Servi-Car first appeared in 1932 and – astonishingly – remained in production until 1973. This is a 1955 model.

OVERLEAF: The future arrives in the shape of the Knucklehead.

SMALLER BIKES

CATCH 'EM YOUNG

Photographs emerged in 2020 of a 338cc Harley-Davidson prototype and much was made of the giant shift the company was planning. As a manufacturer famous for huge, thudding V-twins, it seemed shocking that it was contemplating building such a small capacity bike. As it turned out, the project stalled thanks to a new strategy to concentrate on Harley's core markets, but even if the small bike had made it to showrooms we shouldn't have been surprised – Harley previously spent years making lightweights.

Of course, many of the earliest single-cylinder machines were small capacity – from the very first 106cc powered bicycle to the 493cc Model C of 1929. But these were part of an evolution, a logical progression from the first bicycle with a motor to the world-renowned big V-twins. More interesting is the period after World War II, when as well as developing the famous twins, Harley management decided to diversify into all manner of lightweight sectors – from scooters to two-stroke commuters to minibikes.

From a modern Harley fan's perspective, this might seem curious to say the least, but there was a logic to

OPPOSITE: Young and carefree... you can see what Harley's 1970s advertising was aiming at. But the 50cc M50 shown here never caught on in the USA.

it. After World War II, the American economy surged, thousands of GIs returned home and there was a baby boom of unprecedented proportions – between 1946 and 1964 Americans had approximately 76 million children. Harley realised two things: firstly, a lot of these GIs would need affordable transport; and secondly, that as all those kids grew up, they might like small Harley-Davidsons to get their first taste of freedom. And of course, once won over by the Harley family, these youngsters would surely progress up the ladder and eventually buy the Big Twins. It was a solid theory. The results though, were mixed.

1948 MODEL S

'America's newest sensation on wheels!' is how Harley's 1948 advertisements pitched the new Model S. As ever, the ads were stretching things a tad, but the new bike was certainly popular. Its engine's design was copied from a 125cc, single-cylinder two-stroke made by German manufacturer DKW – as part of war reparations, Harley were allowed to replicate the design, as were Triumph who put it in their Bantam.

Harley built a lightweight frame around the 3bhp engine, added some girder forks that used rubber bands to provide suspension, and a peanut fuel tank that would later appear on the XLCH Sportster. It was simple, economical – 70mpg was relatively easy to achieve – and keenly priced at $325. In 1948 Harley built 10,117 Model Ss – a third of that year's production – though it's thought only half that number were sold that year.

Harley stuck with the Model S though, adding telescopic forks in 1951 and boring out the engine to 165cc in 1953 and calling it the ST.

1955 HUMMER

With the Model ST becoming ever more premium (it had luxuries including clocks and a tail light), Harley decided to introduce a new base model, the Hummer. This was named after Dean Hummer, Harley's most successful two-stroke salesman, and was extraordinarily sparse. It had no horn, air filter, kickstart rubber, brake light or instruments. In fact, it couldn't have had most of those anyway, because it didn't have a battery. As standard it didn't come with any footpegs either, which seems like taking cost-cutting a step too far.

BELOW: The 1955 Hummer used a 125cc two-stroke engine derived from the German company DKW's design.

ABOVE: By the time the Topper arrived in 1960, the demand for scooters was on the wane, particularly for ones started by pulling a rope...

1960 TOPPER

With the influx of Italian scooters made by Vespa and Lambretta in the 1950s, Harley spotted an opportunity to deploy the 9bhp, 165cc two-stroke motor from the Model ST. The Topper was Harley's only attempt at a scooter and was one of their most curious machines – you started it by pulling a rope, like a lawnmower. The engineers laid the ST motor horizontally under the scooter's floorboards to keep the seat height low and developed an automatic three-speed gearbox to make it easier for novices to ride. In some American states you could ride the 5bhp version on a car licence with no motorcycle training needed. The engine was rubber-mounted to reduce vibration, there was a parking brake, and of course plenty of storage under the seat.

Initial sales of 4,000 in 1960 were promising, but Harley was late to the scooter party and it wasn't long before everyone else was packing up and heading home. Scooters were out of fashion and sales declined sharply – Harley sold just 500 Toppers in 1965.

1960 BTH SCAT

As scooters disappeared, so off-road (and off-road-styled) bikes became more popular, and Harley spied another opportunity to deploy the ST engine. The Scat, along with its stablemates the Pacer and Ranger, was another attempt by Harley to broaden its appeal and pick up customers in markets far removed from big twins. All three were powered by versions of the two-stroke single-cylinder used in the Topper, though in the Scat it was bored out to 175cc.

Harley gave the Scat dual-purpose credentials by fitting it with knobbly tyres and fenders with plenty of clearance so as not to get clogged up with mud. But the Scat wasn't a competition off-roader – Harley's advertisements aimed it at the hunting and fishing brigade who needed rugged transport off the beaten track. Initially the bike had a rigid rear end – not optimal when bouncing down muddy tracks – but this was corrected in 1963 with 'Glide-Ride' rear suspension.

The BT Pacer was essentially a Scat aimed purely at road riders, while the Ranger was even more off-road biased than the Scat, with no fenders or lights. All three were eventually replaced by the dual-purpose Bobcat, which was designed by the grandson of Harley founder William A Davidson. It was one of Willie G Davidson's first ever design projects for the company.

1961 MODEL C SPRINT

By 1960 America was awash with lightweight bikes from Europe and Japan featuring high-tech novelties such as direct oil injection (no need to measure out your two-stroke oil) and electric starters – neither of which featured on Harleys. Rather than invest millions in catching up, Harley took the pragmatic approach and bought 49 per cent of Italian company Aermacchi. The Sprint was the first product of their collaboration.

Compared to Harley's standard fare it was fiendishly high-tech. The 250cc four-stroke motor had a powerful overhead cam four-stroke engine made largely of aluminium alloy, while the spine frame, telescopic forks and rear suspension provided excellent handling. In fact, the bike was so capable that in racing fettle it came third in the 1966 350cc World Championship.

However, though unquestionably a fine motorcycle, the Sprint never caught the American public's imagination in the way Harley's big twins did. The fact that you had to rev the road bike's motor to 7,500rpm to extract its 18bhp put off many buyers used to the easy torque of a chugging twin.

Harley's next idea was to put the engine in an off-road chassis and add a high compression piston for good measure. The Sprint Model H – later called the Scrambler – was soon selling more than the road-going Model C thanks to its popularity with flat track and motocross racers. These customers weren't bothered about having to rev the motor hard provided the power was delivered eventually. Which it was.

After 13 years in the line-up, the Sprint's downfall was nothing to do with its engineering and everything to do with economics – by 1974, Japanese rivals were easily as good as the Harley, but considerably cheaper...

ABOVE: The first 123cc Rapidos – like this 1968 model – were designed for the street. Later models had off-road pretensions.

1965 MODEL M-50

Yes, Harley once made a 50cc motorcycle. Once again, the M-50 was an attempt to find sales away from Harley's heartland, and as with so many of the company's other lunges into non-V-twin territory, it wasn't a great success. The engine was a 2.5bhp two-stroke single with a three-speed hand-shifted transmission and was made in the Aermacchi-Harley-Davidson factory in Varese, Italy (modern MV Agustas are made on the same site now). Weighing in at 103lb – less than the engine of a modern Harley Big Twin – the little M-50 was expected to sell in huge numbers so Harley cranked up production and made

more than 25,000 during 1965 and 1966. The trouble was that sales were sluggish, leaving Harley with mountains of unsold bikes that had to be discounted to get them out of showrooms.

1968 MODEL ML RAPIDO

Say what you like about Harley's tactics in the 1960s, but it was nothing if not determined and 1968 saw another bike launched into the cut-throat lightweight market. The 125cc two-stroke Rapido started off as a pure road bike with a four-speed transmission, then morphed into an off-roader. They were simple machines, with the electrical system only powering

OVERLEAF: The SX250 proved to be a popular off-roader. This 1974 model is still going strong, pictured here on the Wall of Death during the Republic of Texas Motorcycle Rally in 2015.

Touch. And go.

Full electrics. The Harley-Davidson SS-350. Takes the kick out of starting. Puts it in the going.

Motivated by the 350cc Harley-Davidson power plant that moves you down express lanes, keeps you hanging in there on back road turns.

All the torque you'll ever need. Neatly ratioed through the 5-speed box.

(With 5-under-foot, who wants 4-on-the-floor?)

A most responsive, beautiful touch. In handling. And styling.

The Harley-Davidson SS-350.

Sets you free.

Harley-Davidson,
Milwaukee, Wisconsin 53201
Member Motorcycle Industry Council

AMF
Harley-Davidson

Harley-Davidson SS-350.
The Great American Freedom Machine.

a headlight and tail light – there were no instruments – and the motor bore a striking resemblance to Yamaha's DT125. The Rapido later spawned the TX, SS and SX models, which all had direct oil injection, but concerns were already being raised about the environmental impact of smokey two-strokes. The off-road SX was selling well, so Harley persisted with the range until 1977, when the company started trying to extricate itself from the Aermacchi deal.

OPPOSITE: By 1973, the Sprint had morphed into the SS350, proudly boasting an electric start. This ad is from *Playboy* magazine.

1972 MC-65 SHORTSTER

No, really: a Harley minibike. Like the Topper scooter and 50cc M-50, the Shortster (like a Sportster but short, see?) is yet another machine that seems at odds with Harley's current image. But the owners of the company were all direct relations of the founders and had a deep understanding of Harley's history. They knew that the early bikes were small, affordable, utilitarian machines, so why not return to making some more?

And the Shortster was a success, selling remarkably well to motorhome owners wanting a small bike to strap to the back of a van. The first machines were powered by a 65cc two-stroke engine, had proper suspension front and rear and were even fitted with a speedo – something lacking from the Rapido models. After a year, the engine grew to 90cc and the bike was disappointingly renamed the X-90. There was even a Z-90 with bigger wheels and off-road pretensions. In total, almost 34,000 of the minibikes were made.

1987 MT350

The relationship with Aermacchi ended in 1978 when Harley sold its share to the Italian Castiglioni brothers, who went on to turn it into Cagiva. After that, Harley steered clear of small capacity bikes until 1987 when an opportunity arose to make the MT350. Bizarrely, this air-cooled single-cylinder trail bike started life with a 500cc Austrian-made Rotax engine and a chassis built by British company CCM. Called the MT500, it ended up being the standard motorcycle for NATO, where it caught the eye of Harley-Davidson, who bought the rights to make it in America. Harley decided to use a smaller engine – a 348cc Rotax with a four-valve overhead camshaft – and created the MT350.

It was a sturdy and well-made machine, though at 162kg it was rather heavy for off-road work – by comparison, Suzuki's DR350 trail bike weighed just 130kg. Still, with 30bhp available and good quality suspension (albeit with twin shock absorbers at the rear rather than the superior mono-shock design) the bike performed well. It even had electric start and front disc brakes – a significant step-up from the old MT500.

Harley sold the MT350 to various militaries around the world, including the British army, and many bikes have made it into civilian life, where the huge panniers, rifle scabbard and khaki paint make it stand out from the usual trail-riding crowd.

Since the demise of the MT350 in 2000, Harley hasn't sold a bike below 500cc, though the company obviously thought about it with the 2020 338cc prototype. As Asian motorcycle markets continue to blossom, it feels like it's only a matter of time before Harley dip another toe into the small-capacity pond.

RIGHT: Until the new Pan America arrived, the Rotax-engined MT was the Harley fan's best bet for trail riding. Front panniers and disc brake indicate this is the 350 and not the 500.

OHV BIG TWINS

THE LEGEND BEGINS

The decision to place an engine's valves on top of the cylinder (an overhead valve, or OHV, layout) rather than on the side (a side valve) might seem relatively trivial – interesting to engineers, perhaps, but of little consequence to the rest of us. But this is most definitely not the case with Harley-Davidson.

When Harley revealed its first ever OHV Big Twin to dealers from around the world at a launch event in Milwaukee on 25 November 1935, it arguably marked the point when the modern Harley legend began. Suddenly Harley had a powerful, beautifully designed V-twin that looked like nothing else.

That first bike was the Model E, also known as the 61 because of its 61 cubic inch (1000cc) engine capacity. The assorted dealers could barely believe their eyes when they saw it – after years of minor model upgrades through the Great Depression, they were expecting more of the same. Instead, they saw innovation everywhere – the chassis, the detailing, and of course that engine, were all new.

OPPOSITE: The Knucklehead was a huge step forward for Harley and the engine was an unqualified success. In 1946 – the year this bike was made – Harley built a record 6,746 Knuckleheads.

ABOVE: So the cylinder heads only vaguely resemble knuckles, but it's a cool name anyway.

The motor was a revelation partly because of the extra power that overhead valves helped to release – it was twice as powerful as the old Flatheads, yet just as reliable (eventually). The motor also had hemispherical combustion chambers, which again helped improve power ('Hemi' heads became famous 20 years later when they were used in American muscle cars). And there was a new four-speed constant-mesh gearbox that was quieter and more reliable than the old three-speed sliding gear version.

Besides the power and technical sophistication, the new motor also looked fabulous. The large, polished rocker covers resembled clenched fists – the engine later became known as the Knucklehead – and pushrod tubes on the right-hand side gleamed. The motor was a work of art.

Then there was the chassis. All previous Harley frames could trace their lineage to the bicycle, with an engine suspended between single tubes. That changed with the 61. Twin downtubes made from strong chromoly steel dropped from the headstock and swept under the engine to the rear axle, where more tubes headed back to the steering head in a straight line. The design was simple, far stronger than any before, and the proportions were near-enough perfect. From the side, you can see the shape of the frame mimics that of the teardrop tanks.

And sitting centrally on those twin fuel tanks was a new instrument panel, with a 100mph speedo, an ammeter, an oil

BELOW: A 1936 EL Knucklehead. Note how the line of the tank flows down through the rigid rear frame to the axle. A design classic, no question.

pressure indicator and the ignition switch. Neat touches were everywhere – from the chrome-plated filler caps to the slash cut air intake. And all that from a company with no styling department.

However, it wasn't all plain sailing for the 61. After Harley showed its most important dealers the bike in November 1935, there was radio silence until February the next year. It turned out that Bill Harley, the engine's designer, had made a mistake with a new oil recirculating system, and while the dealers were clamouring for stock, the engineers were doing a hasty redesign.

With the immediate problems fixed (though there were more to come…), Harley officially launched the 61 on 21 February 1936. There were two versions: the standard Model E and the more powerful EL. By the following month, orders were coming in so fast that Harley couldn't keep up – an astonishing level of demand considering that the US was still recovering from the Great Depression.

Unfortunately, many of the new owners had problems. Though the engine's fundamental design was excellent, Harley had rushed it to market and hadn't cured all the niggling faults that drive owners mad. Those early buyers were effectively riding prototypes and doing work that should have been completed by Harley's test riders. During the first year of production Bill Harley modified almost every engine component.

But customers kept buying 61s, despite all the aggravation, because the performance was exceptional. In stock trim riders would easily see 95mph, which the 61s could hold all day – not something you'd want to try on a Flathead.

By 1937, the 61's multitude of issues had been cured and Harley was ready to reap the sales benefits. The manufacturer's mission was assisted by legendary Harley racer Joe Petrali, who

OPPOSITE: Arthur and Walter Davidson watch the first Knucklehead come off the line in 1936.

ABOVE: Harley was so keen to get NYPD officers on Harleys (instead of Indians) that it built this, the UMG. It was based around a 74 Flathead.

used a tuned Knucklehead to crack 136mph on Daytona Beach in 1936 – a record that still stands. Even more impressively, motorcycle cop Fred Ham rode round a five-mile lap on a dry lake continuously on his completely stock 61 for 24 hours, clocking up 1,825 miles and averaging 77mph.

Over the next few years, Bill Harley and his team of engineers honed the Knucklehead so that by the outbreak of World War II it was outselling the Flathead. In 1941 Harley increased the 61's bore and stroke to push its capacity up to 74 cubic inches (1213cc) for the new Model FL. Because there was already a 74 Flathead, this engine was known as the 74 Overhead.

By 1942, the US had been drawn into the war and Harley switched to producing WLA 45 Flatheads for military use. Over 80,000 Harleys went to US and Allied forces, although, contrary to Harley's advertisements at the time, only a few ended up seeing combat. Most were used as transport for military police and courier duties.

After the war finished, Harley found itself in the perfect position to recover quickly. Though now ten years old, the Knucklehead was still clearly the best American bike. With thousands of GIs returning home, and pre-war bikes often wrecks after years of neglect, sales boomed – Harley built a record 6,746 Knuckleheads in 1946 and a similar number again in 1947.

ENTER THE PANHEAD

For 1948 Harley chose to make the Knuckleheads more refined, and as part of the model update made the cylinder heads from aluminium alloy rather than iron, as well as changing the rocker cover design – now they looked more like American-style cake

BELOW: In 1948, Harley modified the Knucklehead's cylinder heads and the Panhead was born.

pans than knuckles, and so the Panhead was born. If it sounds like the Panhead was more of an evolution of the Knucklehead than a whole new machine, that's because it was. It's the same story with the Shovelhead, but we'll come to that later.

The Panhead did have some new technology that elevated it from the Knuckleheads though: hydraulic valve lifters. These automatically adjusted valve clearances, eradicating one of the more tiresome servicing jobs, and the Panhead was the first ever motorcycle to use them. Theoretically, the hydraulic system saps a small amount of power, but for most owners the benefits far outweigh that single disadvantage.

Along with a multitude of smaller changes – most concerned with making the motor quieter and more oil-tight – the innovations on the Panhead did the trick, and the bike sold well in 1948.

But problems were brewing. In 1948 America initiated the Marshall Plan to help Europe rebuild and prevent the spread of communism and the impact on Harley was immense. Not only was the factory starved of raw materials, Harley also faced an influx of sporty British bikes made cheap by greatly reduced tariffs.

The obvious reaction would have been to take on the Brits with a light, fast rival, but Harley knew that would take years to develop and instead continued to refine the Panhead before introducing a major redesign with the 1949 Hydra-Glide. This had a Panhead engine, huge telescopic forks (hence the Hydra-Glide name) with twice the wheel travel of the old springer forks plus new styling that was more 1950s than 1930s.

There were two areas where Harley hadn't moved with the times though. One was the gear shift. By 1950 every foreign rival used a foot shift system in conjunction with a hand clutch – just as all motorcycles use today. But the Hydra-Glide still had things round the other way, which meant you had to

Thrilling Sport and Thrifty Transportation

HARLEY-DAVIDSON *HYDRA-GLIDE* ®

Every mile is money saved, every moment a thrill...when you ride this swift, smooth "streamliner"! It takes you comfortably to daily work or distant points...at amazingly low cost. It brings you endless good times at race-meets, tours and other exciting motorcycling fun events. With it you can play a vital part in your local defense program. Easy terms. See your dealer. Mail coupon.

Send me free copy of ENTHUSIAST Magazine filled with motorcycle action pictures and stories; also literature on new models.

HARLEY-DAVIDSON MOTOR CO., Dept. P, Milwaukee 1, Wisconsin

Name...

Address...

City.. State..........................

EALERS: Valuable franchises available for the full line of famous Big Twins and the 125 Model. Write today.

After the Panhead came the Shovelhead, which
powers this 1969 FLH Electra-Glide, famous for
being owned by Marlon Brando.

take your hand off the bar to change gear and make delicate adjustments to the clutch with your boot. Eventually Harley relented with this car-style system and listed a foot shift as an option in 1951.

The other area where Harley lagged was rear suspension, which, astonishingly, didn't make an appearance until the 1958 Duo-Glide. This had a swinging arm attached to a pivot at the back of the engine, and two shock absorbers connected to a rear subframe. Finally, Big Twin buyers could cruise along safe in the knowledge that bumps would be soaked up by suspension rather than transmitted directly to their spines.

The late uptake of rear suspension on Harley's flagship Big Twins gives some indication of the company's conservative attitude towards technological progress in this era. In the early decades, Harley was at, or at least near, the technological cutting edge, but by the 1960s it seemed to be continually playing catch-up. Another example was the electric start, which

ABOVE: Harley model names don't really relate to the bikes, but the Duo-Glide is one exception: it means the bike has suspension at both ends...

OPPOSITE: ...and here's an example, a 1960 Duo-Glide.

ABOVE: In 1965, Harley finally fitted an electric starter, creating the Electra-Glide (this is a customised 1966 model).

by then had become common on foreign machinery, whereas Harley's Big Twins still required an arcane starting procedure culminating in a gargantuan effort to kick the engine over. Though hardcore Harley fans regarded this as a valuable method of keeping women, youths and unskilled riders away from their machinery, it did little for sales, and in 1965 Harley introduced the Electra-Glide – essentially a Duo-Glide with an electric starter.

The downside – aside from the hardcore hating the very concept – was that the bigger battery and starter motor itself added weight to an already heavy motorcycle. New buyers didn't care about that though and sales rocketed. But despite the success, Harley had plans to evolve the Panhead engine once again.

SHOVELHEAD ARRIVES

Like the Panhead, the new 1966 Shovelhead engine was an evolution rather than a revolution, so we're really talking about another update to an engine first seen in 1935. Still, with a new cylinder head and rocker boxes that improved cooling, the new Big Twin helped make 1966 Harley's best ever year with sales of 36,310. A lot of those were down to the new Aermacchi lightweights, but the new Shovel (squint and the new rocker covers looked a bit like the back of a coal shovel) was making a decent contribution.

By 1970 Harley had merged with American Machine and Foundry (AMF) in a desperate effort to get funding for

BELOW: Naturally, when the Shovelhead engine appeared in 1966 it was fitted to the Electra-Glide.

new production equipment and development. At first the merger worked well, and talented stylist Willie G Davidson – grandson of co-founder William A Davidson – finally managed to get his design for the FX Super Glide into production.

This was a key bike for Harley. The idea sounds simple: take a Shovelhead engine and merge it with a Sportster front end. But Willie G's design cleverly bridged the gap between Harley and the chopper riders who until then had seen nothing in the range to interest them. Suddenly Harley had connected with its roots – and found a design star who might be able to lead them to further success.

As the relationship with AMF slowly disintegrated and the company was beset by strikes and horrific quality problems, new Shovels continued to appear. In 1977 Willie G's fantastic-looking FXS Low Rider appeared, then in 1979 came the Fat Bob and a year later the Wide Glide. Steeped in the company's history, Willie G recognised that plenty of customers would rather look backwards than forwards, so he introduced models such as the Heritage and Classic.

The engineering department wasn't completely distracted by the AMF-induced turmoil though. In 1978 the Shovel grew to 80 cubic inches (1311cc) and gained electronic ignition. But everyone knew the Shovel was on its last legs – it was fundamentally a 50-year-old engine design – and in 1984 the Evo engine arrived. It marked the start of a new, and considerably happier, time for Harley.

RIGHT: The Willie G Davidson-designed FXS Low Rider is a handsome brute, but early models were plagued with problems thanks to quality control issues during the AMF era.

THE SPORTSTER

BATTLING THE BRITS

British bikes were in the ascendancy in the early 1950s.
The latest Triumphs and BSAs were lighter, faster, better
handling and more technologically advanced than anything
Harley could muster, and they were selling well.
Harley had to respond and, eventually, came up with
a model so brilliant that it remained the backbone of
Harley's range for over 60 years: the Sportster.

The Milwaukee engineers didn't get it right first time though.
In 1952 Harley revealed the Model K, which was the template
for the Sportster and had newfangled technology such as
telescopic forks, rear suspension, a hand clutch and foot gear
shift, just like the British bikes. Those innovations were a step
in the right direction but the Model K's engine was anything
but modern and let the bike down badly.

Whereas most of the British machines had cutting-edge
overhead cam engines, the Model K was lumbered with a
side-valve motor that could trace its design back to 1929; it was
basically an old Flathead 45 with the gearbox mounted within
the engine – an arrangement known as unit construction.

OPPOSITE: The XLH Sportster was a higher compression touring version
of the original 1957 bike. This beautifully restored 1958 XLH is in the
Petersen Museum in Los Angeles.

BELOW: That's more like it. After the disappointing Model K came the XL Sportster with overhead valves, a shorter stroke and more power.

Predictably, the bike did not sell well and Harley hurriedly worked to improve it. In 1954 the engine's stroke was increased, which boosted capacity to 883cc and improved the power, but it was still comprehensively outgunned by the pesky Brits.

Harley knew it had to do more than just tweak its old Flathead motor and in 1957 revealed the XL Sportster. This was more like it. Not only did the new engine have the

overhead valves that would allow higher power outputs, but Harley's engineers had increased the bore and shortened the stroke. This meant the engine could rev higher, extracting yet more power from the 883cc V-twin.

Alongside all this clear thinking was one oddity: the cylinder head. Though Harley had clearly realised the benefits of using alloy for this part of the engine – increased longevity, better cooling – and fitted alloy heads to the Panhead, the

Sportster's heads were made of iron. Why? The early Panheads did leak, so perhaps the engineers were worried about afflicting the new Sportster with reliability problems. Or perhaps Harley's foundry simply couldn't make enough alloy heads… Harley has never revealed the reason.

Still, the new engine – known as the Ironhead – was extremely promising and sales were good. The styling of the early Sportsters had little of the pared-back genius of later models, but it was still a fine-looking motorcycle and appealed to riders who couldn't afford the Big Twins or wanted something lighter and more manoeuvrable.

While XLs poured out of showrooms, Harley's race department was busy seeing what it could make of the new Ironhead engine. The improvements led to the XLH (H for High compression), XLR (R for Racing) and then the XLC (C for Competition), none of which had much of an impact, but are worth mentioning because in 1959 Harley combined the best bits of all three, added lights and a generator and created the road-legal XLCH. Officially the C stood for competition and the H for high compression as usual, but riders christened it the Competition Hot.

Suddenly, Harley had a winner on their hands. Because of the tuned engine, the XLCH was faster than the Brit twins (and almost everything else on the road at the time), and the bike's telescopic forks and rear swingarm suspension meant the handling more or less matched the power. Plus, the peanut tank (small and shaped like, you guessed it, a peanut) made it look like a proper race bike – the XLCH's silhouette set a styling template that Harley would follow for the next 60 years.

With the XLCH selling well, Harley adopted its standard procedure: evolve the bike rather than reinvent it and gradually eliminate the niggles. Starting problems were cured and oil leaks were fixed so that by the 1970s the CH was an

ABOVE: The XLCR
was a sales flop when
it appeared in 1977
– its sporty looks
put off the Harley
hardcore, and its
cruiser-like handling
deterred everyone
else. These days it's a
collector's item.

impressively robust and reliable machine. But, as ever, the world was changing around Harley and the company didn't seem overly keen to react.

The Japanese were launching fearsomely powerful three-cylinder two-strokes as well as four-cylinder four-strokes, both of which were considerably sportier than the Sportster. By 1972 Harley had bored the engine out to 61 cubic inches (1000cc) to gain a few more horsepower and in 1973 the XLCH got a front disc brake. Along with the XLCH's fabulous looks, these changes were enough to keep sales ticking over nicely.

The one element of the Sportster that hadn't changed was the frame, which was heavy and relatively flexible, limiting the bike's handling potential. But help was at hand: Harley's race department was working on a new racer called the XR750) and had designed a far superior frame.

Instead of reinventing the wheel, the road bike R&D department simply swiped the XR750 frame designs, fitted the XLCH engine, took some styling cues from the race bike and in 1977 delivered the XLCR.

This was a huge step for Harley. The brainchild of Willie G Davidson (grandson of company co-founder William Davidson), the XLCR was Harley's first cafe racer, with rearset footpegs, a long reach to the bars, twin disc brakes up front, a single seat and blacked-out styling. It even had a fibreglass nose fairing for riders to tuck behind as they thrashed between bars.

Its fundamental problem was that it turned out to be too radical for the Harley traditionalists to consider buying, yet not sporty enough for someone looking to buy a sports bike – the Japanese and Italian machines of the day were far more powerful and at least 25mph faster. They also went round corners better.

Then there was the build quality. By 1977 Harley was deep into its AMF period with all the ensuing industrial action and upheaval that involved. As a consequence, Harley's reputation for rugged reliability that had been so hard won in the early days started seeping away. Contemporary road tests of the XLCR were largely positive – and amazed that a pushrod engine could compete so effectively with double overhead cam rivals – but were damning about its reliability and build quality. It was also expensive, which didn't help. So, two years after launching the cafe racer, Harley pulled it.

By 1983 AMF was gone, replaced by a group of investors who decided the Sportster range needed a cheaper model at the bottom to tempt people into the Harley fold. The XLX61 – peanut tank, low bars, not much chrome – was priced to match its foreign rivals and was an instant hit.

Also in 1983, Harley launched the XR1000, which was another of the company's imaginative mash-ups of frames and

OVERLEAF: Not all owners wanted their Sportster to be sporty – some just wanted a lighter, cheaper bike than a Big Twin. This 1967 XLH has had the full tourer treatment.

engines from other models. The motor had the alloy cylinder heads from an XR750 race bike, new cylinders and a modified XL bottom end. This was then squeezed into the XLX61 frame. The resultant machine was Harley's fastest ever bike and was tantalisingly close to being an XR750 racer for the road.

But – and you knew there would be one – there were issues. Firstly, all the engine development meant the price was almost double that of the XLX61, yet buyers only got an extra 10bhp over the base bike. Then there was the inevitable problem of foreign rivals being lighter, faster, better handling and, in most cases, cheaper.

In 1984 Harley painted the XR1000 in the traditional racing colours of orange and black so it resembled the XR750, but it was too late. Despite looking fantastic and being a great road bike – the torque at low revs was superb – the XR didn't sell enough. Just 1,777 bikes were built and production stopped in 1984. But, like the XLCR, the XR1000 has attracted more and more admirers as the years have gone by and it's now one of the most sought-after Sportsters ever made.

Things started to turn round in 1986 when the Sportsters got the all new Evo engine that first saw daylight in the Big Twins. Though still a 45-degree V-twin with both pistons connected to a shared crankpin to give the familiar Harley sound and feel, the new engine had an alloy cylinder head. This allowed better cooling, so the engine could be revved harder to make more power, helping inject some sport back into the Sportster.

Fitted with the smallest version of the new engine, the 1986 XLH883 Sportster cost just $3,995 and once again there was a competitive Sportster available at a very competitive price. The bigger 1100cc Sportster wasn't quite such a bargain, but was still $2,000 cheaper than any of Harley's Big Twins.

Realising it had a success on its hands, Harley implemented the tried and tested regime of gradual improvement. In 1992, the 883 got the five-speed gearbox that the 1100 gained two years earlier, which allowed the entry-level bike to cruise more comfortably – the old four-speeder meant the 883 felt increasingly frantic above 70mph. In 1993 Harley replaced the chain final with the belt that had been used on the Big Twins for years. This made the 883 even smoother.

As the smaller Sportster became more refined, the bigger one gained more power. In 1998 the 1200S (S for sport) got a comprehensive engine overhaul that raised torque by an average of 15 per cent across most of the rev range and lifted power close to 75bhp – no great shakes compared to the Italian twins of the day, but enough to make the S feel more sprightly than any Sportster so far. Coupled with fully adjustable suspension (a first for Harley), here was a Sportster that justified the name.

OVERLEAF: The 883 Sportster got the Evo engine in 1986, giving it more power. Crucially, the price was still competitive, providing an affordable route to Harley ownership.

ABOVE: With powerful brakes, an Evo motor tuned for 91bhp and sophisticated suspension, the XR1200 was a great bike. But it proved to be too much for American buyers.

A decade later Harley lifted the sporting bar higher still with the XR1200, which had an even more highly tuned Evo engine that made a creditable 91bhp and would push the bike to 120mph. Its brakes – four-piston calipers and twin discs at the front – were strong, and it even had an alloy swingarm just like Japanese rockets of the day. It was the closest Harley had ever come to producing a true sports bike.

Curiously, the XR arrived in Europe, Africa and the Middle East a year before America, with Harley saying that these were the key markets for such a sporty bike. But poor sales proved that argument was shaky – at 260kg, the XR was a whopping 60kg heavier than the sports bikes Europeans were addicted to and, not having a fairing, looked nothing like them. In 1999 the XR arrived in America and the reception was little better because – you guessed it – it was too sporty for traditional Harley buyers. Once again Harley had positioned a model between two stools, and the XR fell out of the range five years later. However, just like the XLCR and XR1000 from the 1980s, as soon as the XR disappeared from showrooms, it started to gain a cult status. Today, it's highly sought-after by Harley fans.

After the XR, Harley concentrated more on styling changes, creating Sportsters such as the XL1200X Forty-Eight (peanut tank, fat front tyre), XL1200V Seventy-Two (chopper styling) and XL1200CX Roadster (sportier riding position). But with emissions regulations tightening around the world, it was getting tougher and tougher to keep air-cooled engines like the trusty Evolution in production. Harley could either invest a fortune in the Evo to try and keep it legal, or ditch it and plough its money into a new water-cooled engine.

In the end, Harley chose the latter – in a big way. Not only did the company develop a water-cooled engine, but also introduced variable valve timing too. The result was a phenomenally powerful V-twin called the Revolution Max which first appeared in the groundbreaking Pan America in 2020. A year later it appeared in the new Sportster S, a 125bhp rocket that comprehensively blasted all previous Sportsters out of the water. In 2022 three air-cooled Sportsters were still sold in America and some other countries, but the writing was on the wall: the future of Sportsters was water-cooled…

RECENT BIG TWINS

DIFFICULT TIMES

To the outside world, Harley looked in a sorry state in the late 1970s. Its bikes still had the charisma and style of earlier days, but they lagged behind the competition in almost every other respect – they were poorly built, technologically backward, often unreliable and generally leaked oil. The owners of the company – AMF – had the money to keep Harley going, but the management style caused no end of problems with the workforce.

However, behind the scenes, all was not lost. A tight-knit bunch of Harley managers and engineers had a plan: they wanted to develop not one, but two new engines. One would be an evolution of the Shovelhead that was still in production, giving it more power and – crucially – making it oil-tight and reliable. The other was intended to be Harley's long-term future: it would be a cutting-edge water-cooled motor along the lines of those made by Harley's rivals.

Given that most of the Harley engineers had precisely zero experience developing water-cooled motors, the company decided to outsource the project – called Nova – to German

OPPOSITE: This beautiful FLSTN Softail Deluxe looks like it's straight from the 1950s, yet it was made in 2006 and features Harley's innovative Twin Cam engine and Softail rear suspension.

BELOW: The Deuce
was an offshoot
of the Softail
introduced in 1999.
Note the pulled
back bars, stretched
tank and chromed
fork sliders.

engineering powerhouse, Porsche. Hardcore Harley fans were shocked by this decision as details of project Nova leaked out, but it was clearly the right thing to do. At the time Harley did not have the engineering resources to complete the two projects in parallel, never mind one starting from a completely clean sheet.

And the Shovelhead evolution motor – officially called the V2 Evolution, or Evo for short – was a big enough task in itself. You'd think that improving an old engine would be relatively easy – just identify the faults and fix them while introducing a few new bits of technology to keep pace with rivals. The

problem Harley faced was that every major change required tens of thousands of hours of testing to ensure it didn't cause reliability problems, and if it did, then it was back to the drawing board and the process started again.

When you're modifying almost every component in the engine, that time adds up. Harley engineer Hank Hubbard started work on the Evo in November 1977 and the engine wasn't unveiled until late in 1983, though to be fair to the development team it wasn't just engineering problems they were facing. In 1981 Harley sales were the lowest for a decade and motorcycle sales in general were plummeting. Harley's senior managers desperately tried to rein in costs across the board by laying off shop-floor workers and engineers, halving production and cutting the salaries of those who remained. Pushing forward with Evo development under those circumstances was a struggle, but it was worse for the Nova project – that was canned to save money, despite it producing an array of fascinating engines including a V4 that could have revolutionised the company.

But, as is so often the case with Harley, nightmarish times brought out the best in its people, and when the Evo motor arrived in 1984 it was a revelation. The 1340cc V-twin was 9kg lighter and 10 per cent more powerful than the Shovelhead, more reliable, far more oil-tight and yet still looked, sounded and felt like a traditional Harley motor.

At first, the new engine was available in three models – the FLT, FLHT and FXR – all with a new rubber-mounted chassis to take the edge off vibrations. These were an instant hit in showrooms and would have drastically improved Harley's situation were they the only launch. But there was another new bike in 1984; the FXST Softail.

This was a stroke of genius. Realising that many buyers yearned for the clean lines and classic look of the old hard tail

ABOVE: The beauty of the Softail design is that the rear suspension is so well hidden. Early bikes had two shocks slung low while later bikes, like this 2013 model, have one under the seat.

bikes (meaning no rear suspension), but couldn't tolerate the spine-pummelling ride, an independent engineer called Bill Davis came up with a clever solution. By creating a triangular rear swinging arm that joined the main frame at a pivot point just below the seat, Davis was able to hide the rear shock under the engine. Unless you peered hard, the bike looked exactly like a hard tail – clean, elegant and effortlessly stylish – yet had more suspension travel than any Harley at the time.

It was an inspired piece of work and Harley knew it, eventually buying the design from Davis and remodelling it to work with a Harley frame. For buyers who were more interested in style than performance, and couldn't be bothered spending half the time fixing their bike or mopping up the dropped oil as they had to with the Shovelhead, it was perfect. And, because of the Evo engine, their stylish machine was actually surprisingly quick in a straight line, though you wouldn't want to hustle one down a canyon road thanks to a lot of weight and soft suspension.

The following year Harley revealed another clever addition to the range: a rubber belt. Traditional final drive chains are unbeatable if you want a very efficient way to transfer power and don't mind keeping them in fine fettle. But a belt, though marginally less efficient, feels smoother and – crucially – needs barely any maintenance. And it doesn't spray chain lube all over your shiny rear wheel.

So by 1985 Harley had gone from having a moribund range of ancient models to a widely respected assortment of big-selling bikes, many with unique selling points. By 1988 the US motorcycle market was down by 28 per cent, yet Harley had its

BELOW: Harley took the classic look one step further in 1988 with the Springer Softail. Instead of telescopic forks it had old-style girders.

best year for almost a decade, and unveiled yet another stellar motorcycle: the FXSTS Springer Softail.

This had the clever Softail rear end, the new Evo engine and a springer front end that looked for all the world like it had come off a 1940s Panhead. Actually, Harley had spent years modernising and testing the girder-style front suspension system until the engineers were satisfied the bearings would last considerably longer than they did on the original bikes. Again, the designers had got the customer mood spot on: the Springer Softail was an instant hit, despite being one of the more expensive bikes in the range.

Harley was on a roll. In 1990 the FLSTF Fat Boy arrived, with a Softail rear end, solid disc wheels, huge fuel tank

BELOW: With solid disc wheels and shotgun exhausts it has to be a Fat Boy, which first appeared as a prototype in 1989.

and the Evo engine. The combination gave the Fat Boy
a handsome, robust look that, once again, was a hit with
buyers. And the name? Rumours swirled that it was a crass
dig at Japanese competitors, being made up of parts of the
code names for the bombs that were dropped on Nagasaki
and Hiroshima – 'Fat Man' and 'Little Boy'. However,
Harley's vice president of styling and product development
Scott Miller said in 2015 that it was simply down to that
large tank making the bike look bigger than other models
from head-on.

With the Fat Boy selling well, Harley continued to pump
out new models. In 1991 the limited edition Dyna Glide
Sturgis appeared, which at first glance seemed to be nothing
special – just another Harley neatly mimicking the design
of a previous generation (in this case the 1981 FXB Sturgis)
while incorporating some of the company's latest technology.

But there was more to the Dyna than met the eye. The
engine was isolated from the new frame by two composite
rubber blocks that cleverly reduced both the amount of
shudder the rider felt at low revs, and the higher frequency
vibrations at higher revs. In a stroke, Harley had enabled
riders to cruise around town without their fillings being
shaken out, and to comfortably go faster too.

Customers loved the Dyna Glide Sturgis and Harley
soon brought out more models along similar lines. With the
Sportsters fulfilling the role as lighter, sportier entry-level
bikes and the Big Twin tourers at the top of the range, the
Dyna Glides' combination of Big Twin engine and a lighter
and more nimble chassis fitted perfectly in the middle.

So successful were the Dynas that by 1995 they had
replaced all the FXR models, and by 1998 Harley's sales
overtook Honda's for the first time since the 1960s. However,
despite the bikes selling well, the Evo engine was starting

to feel its age and Harley engineers were working hard on its replacement: the Twin Cam.

In many ways this was a bigger deal than the Evo because it was as close to clean sheet design as Harley had ever come. Sure, it was still an air-cooled pushrod V-twin with 45 degrees between the cylinders and both con-rods attached to one crankpin, but its execution was very different from the Evo.

Cranks, crankcases, con-rods, pistons, oil pumps, oil routing, fasteners, capacity… almost everything changed, though plenty of components were designed to externally

resemble the Evo so as to keep hardcore Harley fans sweet. Of course, the crucial difference was the number of camshafts: two instead of one. This allowed the pushrods to line up with the valve gear with less of an angle and therefore run more efficiently, and the chain drive (rather than gears) kept engine noise down to meet new regulations.

Having learned its lesson bringing out motors that were not reliable from the off, Harley worked tirelessly to make the Twin Cam as solid as the Evo, even going so far as using oil jets to spray on the aluminium pistons to stop them

ABOVE: For stylish touring there are few better motorcycles than the Road King. The screen is removable so it can quickly transform into a cruiser.

overheating, then having to increase the size of the fins to try and keep oil temperatures down.

Almost six years after discussions about the Twin Cam first started, the first engines came off the production line in 1998, to be unveiled in bikes the following year. As with the Evo, all the effort was worth it and buyers loved the increase in performance, reduced maintenance and even greater reliability.

To start with the engines didn't appear in the Softails because their frame was too tight for the composite rubber blocks the other Big Twin models used to dampen vibration. And without the blocks, the Twin Cam's vibes were enough to damage the Softail's frame (and reduce riders to jelly). Of course, Harley knew about this, and had a plan. Running parallel to the Twin Cam's development was a project to fit the Twin Cam with two chain-driven balancer shafts to cancel out the vibrations. Once this engine – called the Twin Cam 96B – was ready, it was fitted in the Softails and, yet again, Harley had another sales success in the making...

OPPOSITE: There is no finer way to tour America than from the saddle of a Road King – just sit back and watch the world unfold while listening to the backing track of a vast V-twin.

OVERLEAF: The Softail Springer was made from 1988 until 2003 and were expensive when new because of those complex forks. They're sought-after today.

SPORTING PROWESS

A HISTORY OF WINNING

These days virtually all forms of bike sport – MotoGP, World Superbikes, supercross, motocross, trials – are dominated by Japanese and European manufacturers. Even Harley's previous sporting stronghold of American flat track has been overrun by arch rival Indian, leaving Harley to take consolation in winning minor championships such as the American Bagger series.

But it wasn't always like this. Throughout its history, racing has played an important part in Harley's success, and the company has produced dozens of fabulous racing machines, from early board racers to all-conquering XR750 dirt trackers to Grand Prix-winning short circuit bikes. Though racing might seem a long way from Harley's current model range, its sporting heritage runs through the company like the writing in a stick of rock.

In the early days, winning reliability trials was a key way to prove how robust the new Harleys were, and co-founder Walter Davidson gained valuable publicity for scoring a perfect 1,000 points in a 1908 trial on a stock Model 4. It was a good

OPPOSITE: Sam Arena won the Pacific Coast Championship in 1938 by keeping his Harley WLDR at full throttle for the entire 200-mile race – a major feat in those days.

start, but the reliability trials were soon overtaken by a far more exciting pastime: board tracking.

These races took place around huge banked ovals resembling overgrown walls of death constructed from planks of wood. Spectators crowded round the edges to watch the bikes shoot past at speeds over 100mph, and couldn't fail to be impressed by whichever manufacturer built the winning machines. And if it sounds dangerous, it was – the motordromes where the races were held were dubbed 'murder dromes' because of the frequency with which racers and spectators perished.

Despite the carnage, their popularity grew rapidly from 1915, and Harley worked hard to create a winning bike. William Harley was in charge of the factory team and took a pragmatic approach, hiring outside experts to help get more

BELOW: Famed Harley racer Joe Petrali was the US hill climbing champion in 1932, 1933, 1935 and 1936.

OPPOSITE: It's not recorded where or when this took place, but it's a fair guess it didn't end well. The bike is a modified Model F.

power and better handling from the 999cc V-twin Model 17s. His efforts were successful: in 1921 Otto Walker won the Fresno board mile at an average speed of 101.4mph – the first time any motorcyclist had won a race averaging over 100mph. Harley's team of riders – called the Wrecking Crew – won all the national titles that year.

Thanks to the Great Depression, racing took a back seat in the 1930s as both Harley and its customers hunkered down to weather the storm. And as racers emerged from the financial wastelands, they discovered the racing landscape had changed.

The huge board tracks were largely gone, replaced by a variety of dirt tracks known as TTs.

In 1941 Harley revealed two new race bikes, the WR and WRTT, based on the Model W and powered by the 45 cubic inch (740cc) Flathead engine that first appeared in the Model D of 1929. The WR had a lighter frame designed for oval racing, while the WRTT's frame was built to survive heavy landings from the jumps common on TT courses.

Though purpose-built racers with no lights, speedos or mudguards, these were not sophisticated machines. The WR had basic girder forks, no brakes, no rear suspension and a hand change gearbox, while the WRTT was only a little better equipped – it used the road bike's wheels and brakes.

BELOW: A 1930s press shot of Australian speedway rider Don McPherson on a 350cc Peashooter. Note the lack of drive-chain!

However, the bikes were tough and popular which meant they often won races by either outlasting rivals or benefiting from a rider skilled enough to overcome their deficiencies. In 1948 the WRs won 19 out of 23 Class C championship races – the category for road bike-based racers.

However, good though the WRs were, they weren't racing on a completely level playing field. The American Motorcyclist Association (AMA), which organised the races, was very keen for US manufacturers to do well, and introduced rules to limit foreign bikes to 500cc, while the Harleys were allowed up to 750cc.

Though these rules helped keep the WRs in contention, by the 1950s it was clear the bikes were on borrowed time. However, those rules meant Harley didn't have to innovate to stay in contention, and instead of coming up with a cutting-edge race bike it launched the KR and KRTT.

These were based on the Model K, which had a hand clutch, foot shift and rear suspension but was afflicted with an antiquated side-valve engine. The race engines got upgrades such as needle bearings and more aggressive cam timing, but they were fundamentally from a different era than their foreign rivals.

Still, much like the WRs, the KRs punched well above their technological weight, helped by the AMA's favourable rules. During 18 years of racing they won 12 American national titles, though when the AMA removed the 500cc limit for overhead cam (meaning foreign) bikes, the KRs struggled. Harley knew something had to be done, and what it did would result in one of the greatest race bikes of all time.

The early signs weren't good though. Harley's race engineers took an old Sportster XLR engine which was built for TT races that allowed overhead valve engines. After tuning the motor, the engineers bolted it into a KRTT frame

and called it the XR750. It was seriously flawed. Because the cylinder head was iron, it got hotter and hotter through races until it eventually melted and the engine blew up.

For the 1972 season, the XR750 had alloy heads and was immediately fast, winning the AMA dirt track title in its first year. On the tarmac circuits the XR usually got toasted by the British overhead cam bikes and new two-strokes from Japan, but on dirt, the Harley's ability to find grip and use all its power still made it the weapon of choice. In fact the British

ABOVE: Cal Rayborn and his XR750 shocked the British in the 1972 Transatlantic races, winning three times despite never having seen the UK tracks before.

ABOVE: Still going strong... this is a 1974 RR250 being put through its paces in 2021.

marques got so fed up of losing they withdrew from dirt track racing altogether.

However, the XR wasn't unbeatable. In 1973 and 1974 Kenny Roberts took the title on an underpowered Yamaha XS650, and he famously used a ludicrously powerful Yamaha TZ750 to win the Indy Mile in 1975. When Yamaha withdrew from dirt track in 1977 to concentrate on road racing, the field was once again clear for the XR750 to win everything.

The 1970s was a fertile time for Harley away from the dirt too thanks to its purchase of Italian firm Aermacchi. The Italian engineers were at the forefront of two-stroke engine design (one of the reasons Harley bought the company) and their twin cylinder race bikes were the equal of anything made by the Japanese. By Harley's standards, the performance was other-worldly, with the 250cc motors making over 50bhp, and the 350s – which were 250s with bigger bores and longer strokes – producing close to 70bhp.

In 1971 Italian racer Renzo Pasolini took three Grand Prix wins on the Harley RR250 and won the Italian championship too. It was a sign of things to come: the following year he finished second in the 250 GP world championship and third in the 350. He was all set to go one better in 1973, but was tragically killed in the Italian GP, and his place was taken by fellow Italian Walter Villa.

Villa dominated 250 GPs on the Harley RR250 for the next three seasons, winning each championship with relative ease. He also won the 350cc world title in 1976 on the 350cc version, by then fitted with hydraulic disc brakes instead of the 250's drums. The Harleys were successful over the next few years – Franco Uncini and Walter Villa won five GPs on the RR250

BELOW: The Aermacchi-built RR250 was a brilliant race bike, winning the 250 Grand Prix championship in 1974, '75 and '76. Not bad for two 125cc two-stroke motocross engines joined together...

ABOVE: Plenty of
XR750s are still raced
in America, though
such is the model's
fame that many
spend their lives on
parade. This one is
being ridden at the
GoodwoodFestival of
Speed in the UK.

OPPOSITE: Harley
factory rider Randy
Goss (left) won two
Grand National
titles on the XR750.
His team-mate
Jay Springsteen
won three.

in 1977 – but those were the last Harley GP victories. In 1978
Aermacchi was sold and Harley concentrated on its four-stroke
V-twin racers.

There seemed little to worry about on the dirt ovals
where the XR750 still reigned supreme in the early 1980s.
Then, in 1984, Honda entered the fray and things got more
troublesome. Instead of reinventing the wheel, the vast Japanese
company bought an XR750, analysed each component and
improved every part it could. The result was a machine with
a four-valve per cylinder overhead cam V-twin that was more
powerful than the XR, yet had a similar ability to find traction.
Honda's RS750 won the championship in 1984, '85, '86 and
'87. Fed up with watching a Honda win, the AMA decided to
give Harley a break and introduced intake restrictions to slow
the RS750 down. Honda promptly left the series and Harley
XR750s climbed back on to the top step of podiums again
(well, apart from in 1993 when a privately entered RS750
won the championship). For 21 years from 1994 until 2015, a

Harley XR750 won every single championship, creating stars such as Scott Parker and Chris Carr.

On the tarmac circuits, meanwhile, Harley got the opportunity to take some wins in 1983 when the Battle of the Twins series was introduced to try and get American fans back into racing. It worked a treat, as a Harley XR1000 and dirt-tracker-turned-road-racer Jay Springsteen racked up wins. Though the racer was used to publicise the new XR1000 road bike, the engine was anything but standard, consisting of an XR750 bottom end, alloy cylinder heads, two spark plugs per cylinder and twin smoothbore Mikuni carburettors. The result was 106bhp at 7,500rpm, though the team reduced the rev ceiling to improve reliability so Springsteen would have had just over 100bhp available. Nicknamed Lucifer's Hammer by Harley's race chief Dick O'Brien, the bike ruled the Battle of the Twins for three years until the series was swallowed up into the new Formula One class.

Eventually this turned into the Superbike class, where bikes with different engine configurations could fight it out thanks to rules that kept the playing field more or less level. In the late 1980s, the popularity of Superbike racing was soaring, with fans hooked on watching 888cc V-twin Ducatis dicing with 750cc four-cylinder Japanese bikes. Harley saw an opportunity: if the company could build a competitive V-twin race bike, that could help change the growing perception that Harleys were stodgy cruisers suitable only for old men.

And so the VR1000 project was born. When Harley announced that it would build the entire machine in-house, many commentators feared the worst – how could a race department specialising in V-twin dirt trackers build something that would be competitive in the AMA superbike series? It would, after all, be up against the might of factories which had spent years honing their machines.

OPPOSITE: Between 1988 and 1998 Scott Parker won nine Grand National flat-track championships for Harley. Here, he's just won the race at Daytona in 1990.

RIGHT: To qualify for AMA racing, which is for production bikes, Harley had to build 50 road-going VR1000s in 1994. The AMA didn't say where the bikes had to be legal on the road though, so Harley chose Poland…

But Harley knew the scale of the challenge and didn't hold back. Yes the VR1000 was a V-twin, but the angle between the cylinders was 60 degrees not 45, there were twin overhead camshafts operating four valves per cylinder, it was water-cooled and fuel-injected, and had an alloy beam frame. Stripped of its carbon fibre fairings, the bike looked as lithe and purposeful as any Ducati. When the bike was first rolled out in 1994, it wasn't just traditional Harley fans whose jaws hit the floor.

The first VRs produced 120bhp at 10,800rpm which made them fast, but not quite fast enough – some of the Ducatis made 20bhp more. Still, the VRs handled well and with top-level racers on the bikes they did wellt. But Harley wanted to win, and that wasn't happening.

By 1999, a comprehensive engine overhaul had produced massive power gains – up to 170bhp – and results improved, but the VRs still never got to the chequered flag first. Eventually, Harley management had had enough and canned the project in 2001. The company hasn't returned to tarmac racing since.

However, all that time, money and effort spent on the VR1000 wasn't a complete waste. Harley's engineers learned a huge amount from developing the VR1000 engine and a lot of that insight went into a new water-cooled, 60-degree V-twin road bike: the V-Rod. And, 11 years after the V-Rod appeared, that same format was used in Harley's most radical road bike of modern times, the hugely successful Pan America…

BRAND POWER

HARLEY *IS* MOTORCYCLING

Harley's status as a world-famous brand is unquestioned.
Not only is Harley the most recognised motorcycle
manufacturer on the planet, but the impact of those bikes and
their riders on society has broadened Harley's influence far
beyond the world of two wheels. From the choppers starring
alongside Peter Fonda and Dennis Hopper in *Easy Rider* to the
Fat Boy ridden by Arnold Schwarzenegger in *Terminator 2*,
Harleys have managed to permeate modern culture like
few other products.

And the connection Harley has with its hardcore buyers is
so intense that it has been studied by branding experts for
decades. Loyalty to a brand is one thing, but having the
company logo tattooed on your body – as many Harley owners
do – takes things to a whole new level.

But it wasn't always like this. Modern riders are often
surprised to learn that Harley's values of freedom and
individuality, plus the perception among some that riding
a Harley makes you part of a rebellious brotherhood, are a
relatively recent development. For the first 50 years of its

OPPOSITE: The Terminator wasn't going to ride anything else, was he?
Arnold Schwarzenegger aboard the Fat Boy from *Terminator 2*.

existence, Harley's reputation was built solely on its bikes being rugged, reliable and more refined than the competition.

In fact, Harley's first successful V-twin, the 1911 Model 7D, was marketed as 'The Silent Gray Fellow', with advertisements focusing on the engine's mellow exhaust note and the bike's subtle paint scheme. In those days the prime requirement for a motorcycle was that it got you where you wanted without fuss rather than making a statement about your individuality. Though Harley's sporting successes were useful in adding a

dash of glamour and speed even in those early days, much of Harley's pre-World War II advertising concentrated on convincing buyers that its bikes would be powerful enough to whizz you up hills and not break down along the way.

That pattern of brand development continued until World War II ended, when tens of thousands of US soldiers returned from the battlefields of Europe with cash in their pockets and a longing for excitement. Finding civilian life boring and restrictive, thousands of them bought Harleys and spent their

leisure time riding, criss-crossing the country in search of new adventures.

OPPOSITE: Evel Knievel aboard his XR750 in 1977, the year he tried to jump a pool full of sharks and crashed.

One problem they found was that their Harleys were heavy and slow compared to the bikes they had seen or experienced in Europe. Consequently, the ex-soldiers started unbolting all the parts they considered unnecessary and chopping off bits that seemed to add weight and little else. These machines were the first choppers (long forks, no frills) and bobbers (mudguards chopped down, even fewer frills), both of which went on to have a dramatic effect on the perception of Harleys.

Running in parallel with the ex-soldiers' mechanical alterations was a desire to mix with like-minded souls, usually other veterans on Harleys. Most of these groups caused no issues – in fact you could argue they helped the traumatised young men get the support they needed from fellow veterans and slowly reintegrate into society. But several of the groups took an appetite for excitement one step further and became the first biker gangs. Run and staffed by experienced ex-soldiers – some of whom were no doubt suffering from PTSD – the gangs discovered that violence was an easy way to make money. Armed robberies and gun-running quickly led to protection rackets and drug distribution – they became organised criminals, just like the Mafia.

Matters weren't improved by the so-called Hollister riot of 1947, where fighting broke out at a motorcycle rally in Hollister, California and sensational press reports portrayed a scene of lawlessness and debauchery. An image of a motorcyclist sitting on his Harley drinking, surrounded by empty bottles, was splashed across the influential *Life* magazine, reinforcing a public perception that Harley riders were outlaws. Later it became apparent that the 'riot' was nothing more than a pub brawl and that the debauchery largely consisted of bikes being ridden up and down the main street at all hours – annoying, certainly, but not exactly a riot.

Still, the damage was done and by the 1950s, the American public's perception of Harley's motorcycles went from rugged and powerful modes of transport to something altogether more dangerous. Harleys were ridden by people you didn't mess with – bad boys at best, violent criminals at worst. The impression was so strong that even now, if you ask what bike Marlon Brando rode in the biker outlaw film *The Wild One* – inspired by the Hollister riot – most people will say a Harley, when it was actually a Triumph. Though, to be fair, Lee Marvin's gang in the film did ride Harleys.

Of course, Harley had nothing to do with these perceptions and by the mid-1950s had brought out the 125cc Hummers in an attempt to diversify and prove to the public that it was more than merely a manufacturer of massive V-twins. Harley's purchase of Aermacchi in 1960, and the consequent flow of Harley small-capacity bikes, also demonstrated Harley's willingness to diversify.

The problem was that this didn't work – the little bikes didn't sell enough to be a viable long-term proposition, partly because their main rivals were made by the clean-cut Japanese brands that appealed to women and younger people. The company supplying transport to the Hells Angels wasn't a perfect fit for cute scooters. So Harley concentrated on its core market – big V-twins – and instead of fighting the anti-establishment perception, it tried instead to channel the acceptable aspects: machismo, independence and freedom.

This proved to be a master stroke and was so successful that it allowed Harley to ride out periods such as the AMF take-over from 1969 to 1981 when many of the motorcycles were shockingly badly built. A company without such a strong brand image would surely have perished.

Another advantage of the bulletproof brand was that Harley's marketing department could lever the logo to sell other stuff. The first licensing deal was in 1981 when Harley branded beer appeared, followed by everything from aftershave to cake decorating kits. It's a measure of the brand's strength that even the most dubious product associations failed to seriously dent Harley's image. And the brand extensions make a lot of money. In 2021, almost a quarter of the company's

BELOW: The Hells Angels Motorcycle Club was started by a group of WWII veterans. By the time this photo was taken in 1964 it had more sinister overtones.

ABOVE: Mugs, T-shirts, socks, aftershave, beer, toilet seats... if it exists, it's probably had a Harley logo on it at some point.

OPPOSITE: This 1991 action film got mixed reviews but branding experts the world over were swooning at Harley's triumph.

revenue came from licensing deals and sales of accessories and general merchandise.

The 1980s was a boom time for Harley's marketing department. In 1983, just two years after starting the wildly successful brand extension campaign, came another crucial innovation: the Harley Owners Group, or HOG. This was based on the multitude of small clubs that had sprung up over the years and the idea was simple: encourage Harley owners to ride more, meet more Harley riders and become part of the Harley family. Get it right and owners would never stray to other brands because too much of their motorcycle life was built around Harley and HOG.

It worked so brilliantly that there are now over one million HOG members, 1,350 chapters (Harley borrowed the bike gang term for a local group and use gang-style patches on jackets too) and it's by far the largest factory-sponsored

ABOVE: Huge Harley rallies are famous in America but such is the brand's power they happen elsewhere too – this one is in Barcelona, Spain.

motorcycle club in the world. Each chapter – usually based at a Harley dealership – organises rallies, runs and parties, so if you're buying your first Harley, you have the chance to join a ready-made group of like-minded riders that, without HOG, could take years to build. Harley dealers are able to sell you a motorcycle and a ready-made lifestyle. It's no wonder that other manufacturers have blatantly copied the formula, though none have been anywhere near as successful as HOG.

Another area where Harley has led the world is modifying. Of course motorcycles of all brands have been customised by their owners since the dawn of motorcycling, but Harley riders have generally been the most prolific and ambitious, spawning dozens of new subcultures, from choppers to bobbers to flat-trackers.

Sometimes modifications are done to improve the way the bike goes – the motivation of the GIs returning from World War II, for example – but more recently the changes have been less about function and more about form. They're done to create a bike that looks different from everyone else's. After ignoring this phenomenon for decades, in the 1960s Harley started to realise that it would be commercially astute to embrace it and the company started selling customisation parts from dealerships. Eventually Harley made bikes that had a lot of the customisation work done already – the first of these 'factory customs' was the 1970 FX 1200 Super Glide.

Such was the demand for customisation that professional outfits were established to turn riders' wildest visions into

BELOW: Created by legendary custom builder Arlen Ness, this bike is a tribute to the 1957 Chevy Bel Air. The aluminium bodywork took eight months to hand-form.

OVERLEAF: Harley's dealer network in America is vast, as are the dealerships themselves. Each one hosts a branch of the Harley Owners Group (HOG).

reality. Artisans such as Arlen Ness and Ben Hardy created machines — choppers mostly in their early days — that blurred the line between motorcycles and art. Besides jaw-dropping looks, the bikes often featured wildly imaginative engineering too — superchargers were added to motors to boost power and Arlen Ness even joined two Harley engines together at the crank to create a monstrous V4.

Today customisation is a crucial part of the Harley brand, with the company's accessory catalogue regularly exceeding 700 product-packed pages. Bikes like the Sportster 883 and 1200 were actively sold as blank canvases, ready for customers to turn them into their own personal Harley, and when the new Sportster S arrived, there were complaints that it was too sophisticated and difficult to modify.

The new Sportster is part of Harley's latest plan to concentrate on what it calls its 'core strengths' rather than chase new markets. This is a bold move, because since 1985 the average age of a Harley rider has been creeping up – in 1985 it was 35 and it's now estimated to be near 60 (Harley stopped releasing figures in 2008 when the average age was 48). This is a problem because older customers tend to ride fewer miles, buy fewer bikes and parts, and have an annoying habit of stopping riding altogether or indeed dying. The bulk of Harley's recent sales problems have been blamed on this trend.

The previous strategy was called 'More roads to Harley' which tackled the ageing problem head-on. Not only did Harley launch the electric LiveWire under this regime, but the company said it would launch a whole raft of smaller electric bikes to get youngsters on to Harleys and spark a resurgence in company fortunes. And yes, the parallels with the foray into small bikes in the 1950s, '60s and '70s are startling.

However, with sales continuing to slide, a new CEO was brought in and he decided that the 'More roads' strategy wasn't going to work in time, so ditched it. Instead, Harley has split off the LiveWire part of the business and is now solely concentrating on Big Twins once again…

MODERN TIMES:

FROM V-ROD TO PAN AMERICA AND LIVEWIRE

PROGRESS LIKE NEVER BEFORE

Since the year 2000, Harley has experienced one of the most exciting and tumultuous periods in its entire history. In that time the company has launched four radical new motorcycles, bought MV Agusta then sold it, taken over Buell Motorcycles then closed it, suffered huge industrial disputes, welcomed 250,000 fans to its 100-year celebrations in Milwaukee, dramatically changed the company strategy twice and opened a vast new museum.

For those critics who have accused Harley of just plodding along, churning out the same old V-twins, the last couple of decades must have come as something of a shock.

The bike that clearly demonstrated Harley's desire to move forward into the modern era was the V-Rod, which was

OPPOSITE: A queue of electric LiveWires await test riders at a demonstrator day in 2019. The bike was expensive, but also superb.

launched in 2001. Though still a big V-twin, this was a radical departure from Harley's traditional fare. It was liquid-cooled, had no pushrods, used four valves per cylinder head, had a 60-degree V-angle between the cylinders instead of 45 and used the wildly oversquare engine architecture of Japanese and European performance bikes (where the cylinder bore was far greater than the stroke). In fact, it was designed by Europeans – Harley engineers worked with Porsche on the engine, as they had done with the defunct Nova project in the 1970s. The performance was modern too – the 1131cc engine made 115bhp at a stratospheric (for Harley) 8,250rpm and torque was a thumping 84 lb.ft at 7,000rpm.

BELOW: From this angle, the V-Rod looks pure Harley – kicked out forks, low stance, plenty of chrome. But from the front you can see the radiator for the all-new liquid-cooled motor...

Though new for a production bike, the engine (named the Revolution) wasn't new to Harley. The Porsche engineers used the VR1000 race motor as a starting point, and the fundamental architecture of the two engines is identical.

The most significant change from past Harley motors was the Revolution's liquid cooling, which had a huge aesthetic impact. Part of the beauty of Harley's traditional V-twins is their visual simplicity – just two big cylinders, some exhaust pipes, an air cleaner and that's about it. But with a liquid-cooled engine you get a lot more clutter. For a start there's the large radiator needed to cool the liquid itself, and this works best out in the breeze which means it's usually placed in front of the engine. Harley's stylists worked hard to disguise it by creating a shaped aluminium housing, but it still looks clumsy compared to a sparse air-cooled motor.

OPPOSITE AND: BELOW: When the V-Rod was launched in 2001 its engine shocked the motorcycle world: it was smoother, more refined and far more powerful than any previous production Harley.

Then there's all the pipework needed to move the liquid around – take the fairing off a modern Japanese bike and it can look like the back of a washing machine. Here Porsche's engineers had more success, cleverly designing coolant channels into the engine itself to avoid ugly pipes spoiling the look. They even cast the cylinders with fins, to give a faux air-cooled appearance.

Given all these difficulties, why bother with liquid-cooling? The first major reason was noise: with regulations becoming increasingly strict, liquid-cooled engines have an advantage because the noisy combustion chambers and valves are surrounded by water pipes which muffle the racket. Like every other manufacturer, Harley recognised that air-cooled motors will eventually be regulated out of existence.

The other advantage is power: because a liquid will cool an engine more efficiently than air it allows designers to push engine components to greater extremes, and with more revs comes more power, with no loss of reliability. And although Harleys are not bought on the strength of their power alone, getting closer to rivals' figures gives the bikes wider appeal.

Though the V-Rod's engine was a joint venture, the rest of the bike was pure Harley, coming from Willie G Davidson's design studio. He wanted a drag racing vibe to emphasise the engine's high power output and concentrated on making the chassis look long and low. Controversially, he also kicked out the forks for an even more authentic drag racing look. Normally, this lumbers a bike with ponderous handling, but by experimenting with tyre selection and the other aspects of steering geometry, Harley engineers managed to work round the 34-degree rake angle (sports bikes, by comparison, have a rake of 24 degrees).

When the bike arrived in showrooms it was a revelation: smoother and more powerful than any production Harley

ever. Snick through the German-made five-speed gearbox into fifth and you could eventually end up doing 140mph. Motorcycle magazine road testers could barely believe it came from the Milwaukee factory, such was the performance and sophistication.

Sales were buoyant at first, with waiting lists at most dealers, but it was never the rip-roaring success Harley hoped for. New models arrived with added power and some with even more drag-orientated styling – including a humongous 240-section rear wheel – but sales were nothing more than steady and the V-Rod was discontinued in 2017.

ABOVE: The Street 750 was a smooth, good value motorcycle. Its problem was that it didn't look or feel much like a Harley – the company stopped selling it in 2021.

While the V-Rod was changing some riders' perceptions of Harley, the company decided to take things further by buying in some sporting expertise. In 2003 Harley took over Buell, which was run by ex-Harley engineer Erik Buell. The ex-racer specialised in wildly innovative bikes built around Harley XR1000 motors, and the plan was to use him and his team as a 'skunkworks' for developing sportier Harleys.

But that plan changed, and Buell ended up being the 'starter brand' for Harley – the idea was that customers would buy a Buell, then move on to a Harley. Buell made the 492cc single-cylinder Blast which was used in Harley's training schools but unsurprisingly Erik Buell was none too pleased with the arrangement. The relationship deteriorated and in 2009, Harley closed Buell.

This decision was taken by Harley's new CEO Keith Wandell, who made another important call soon after getting his feet under the desk: he sold the famous Italian bike brand MV Agusta, which the previous management team had bought just a year earlier. Wandell was under intense pressure – the banking crisis and resultant economic carnage meant Harley's profits dropped by 84 per cent in 2009.

Wandell decided that some of Harley's problems stemmed from the lack of younger riders being tempted into the fold. He set in motion a project to create an entirely new entry-level bike that would do a similar job to the Buell Blast, but be a 'proper' Harley – a V-twin. The engineers came up with a cunning plan: by using the technology developed f or the V-Rod, they could create not one but two smaller-capacity bikes.

Both the Street 750 and Street 500 were launched in 2013 and powered by mini versions of the V-Rod's Revolution engine, with the 500 just being a sleeved-down version of the 750 (identical in every way, except it had smaller pistons

OPPOSITE: The Street 750 was intended to be aspirational but affordable in India, and to bring younger riders into the Harley fold in Europe and America.

OPPOSITE TOP:
Harley's then
marketing boss
Mark-Hans Richer on
a LiveWire prototype
in 2014. The bike
launched in 2019.

OPPOSITE BOTTOM:
Until the new
Revolution Max bikes,
the Livewire was
Harley's most high-
tech bike ever. Its
battery recharges as
it slows down, giving
it a range of over 140
miles in stop-start
urban conditions.

OVERLEAF: Harley
knew the LiveWire
had to perform
superbly and ooze
build quality. And
it did both. The
downside was
the price.

and cylinders). Called the Revolution X, the new motor shared many fine traits with its originator: smooth, reliable and easy to use. The Streets seemed rather docile and characterless compared to Harley's brawny Big Twins but, hey, they were starter bikes.

There was innovation in the manufacturing process too. Streets for the American market were made in America as normal, but the rest of the world's bikes were produced in Harley's Indian factory. This made lots of sense: the Asian markets were booming, with millions of riders looking to step up from a 125cc bike to a premium machine like a Harley. Meanwhile, the cheaper costs of the Indian factory meant the Streets were great value in Europe, where they were marketed as an entry point to the Harley range.

As with Harley's previous attempts to bring in younger riders (e.g. Hummers, Toppers, Aermacchi Harleys and the Blast) the Street 500 and 750 didn't have the impact Harley wanted. Sales in India were fine to start with, but competition from a rejuvenated Royal Enfield and a desire for more adventure-styled bikes slowed demand. As with the Blast, a new CEO arrived, reviewed the situation and decided to concentrate on the core market. In 2021 Harley announced the Streets would be discontinued and the Indian factory closed down.

A year after the Streets were launched, Harley revealed an even more radical departure from V-twins: an electric motorcycle called the LiveWire. Because it was a prototype, most industry experts dismissed it as a marketing stunt to try and give Harley some eco credentials. At the time, no major motorcycle company had even admitted to developing an electric bike, so the chances of Harley – with its history of thumping V-twins – bringing one to production were regarded as negligible.

But, astonishingly, that's exactly what happened. In 2019, Harley launched the LiveWire and became the first major motorcycle company to have an electric bike in its range. The reviews were rapturous: the LiveWire was as fast as a superbike away from the lights thanks to the electric motor's colossal torque, it handled superbly despite the battery's weight and it was fabulously designed and built.

Of course, as with many electric vehicles, the range was an issue. If you rode gently you might get 70 miles on open roads, but if you pressed on, the battery could last less than 40 miles. However, in urban areas the range was far better thanks to the lower speeds and regenerative braking – you could easily get over 100 miles on a charge.

The main problem was the price: it was £28,750 in the UK and $29,799 in America – enough to put off all but the most enthusiastic (and wealthy) Harley fans. Sales were modest, and in 2021 Harley decided to create a separate business to sell the bikes, marketed under the LiveWire brand. Crucially, these identical machines were nearly $10,000 cheaper than the originals though it's too soon to know whether sales have picked up to a sustainable level.

As if an electric bike wasn't enough, in 2021 Harley landed two more bombshells. Firstly, no more Sportsters were coming to Europe because of problems getting them through emissions tests, and their future in America looked doubtful too. And secondly, Harley launched an adventure bike, not only powered by an all-new water-cooled V-twin making 148bhp, but also festooned with cutting-edge electronic rider aids.

In a way, this bike – called the Pan America – was even more shocking than the LiveWire. The electric machine could be regarded as Harley heading out on a curious limb (so curious, in fact, that it became a different company), whereas the Pan America was Harley going head-to-head with

European and Japanese manufacturers in a sector they had been honing for decades and Harley hadn't set foot in for over 50 years.

The Pan America's new engine, called the Revolution Max, was extraordinary in itself. The 1252cc V-twin had variable valve timing to maximise both mid-range wallop and top-end power, featured a different crankpin offset so it didn't have the traditional Harley 'potato-potato' sound, and even had sodium in the exhaust valves to aid cooling. It was a state of the art engine, not just by Harley's standards, but by any.

The rest of the bike was similarly high-tech. The higher spec of the two launch models came with semi-active suspension,

BELOW: Water-cooled, variable valve timing, multiple modes... the Pan America engine is cutting edge. Its semi-active suspension is equally modern.

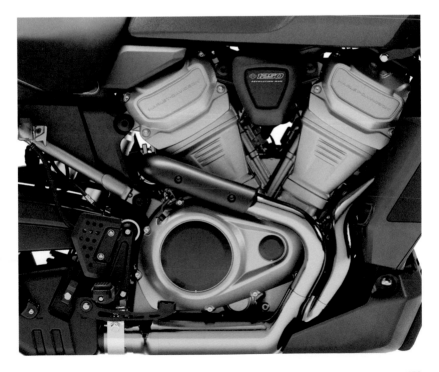

which adjusted its damping depending on your riding style and the road surface, and an innovative system that lowered the bike as it came to a standstill to allow shorter riders to touch the floor more easily. If magazine road testers were awed by the LiveWire, they were flabbergasted by the Pan America, and initial sales were impressive – Harley struggled to make enough of the bikes in 2021.

In 2022 a detuned version of the Revolution Max engine was used to power a new model called the Sportster S. Despite the name, this was more like the old V-Rod than a Sportster, as it was powerful, relatively expensive, looked like a road-going dragster and confronted Japanese and European rivals head-on.

These latest bikes bode well for the future of Harley. The fact that in the last decade the company produced a genuinely competitive adventure bike, a cutting edge V-twin and a superb electric motorcycle has echoes of those early years when William S Harley and Arthur Davidson were furiously innovating in their tiny workshop. If the modern Harley engineers can channel that spirit and determination, the company should be good for another 100 years…

RIGHT: Taking such a huge and expensive motorcycle off-road takes some confidence. But thanks to years of testing in the dirt, the Pan America is remarkably proficient.

INDEX

(Key: Page numbers in **bold** refer to main entries incl. photos/captions, *italic* to all other photos/captions)

CREDITS

The publishers would like to thank the following sources for their kind permission to reproduce the pictures in this book.

ALAMY: Album 129; Alvey & Towers Picture Library 22; George Atsametakis 59; David O. Bailey 90; Neil Baylis 55; Paul Briden 86; Chronicle 12; CJM Photography 110; Michael DeFreitas North America 31, 68-69; Chuck Eckert 21; Entertainment Pictures 120; Bernie Epstein 49; EuroStyle Graphics 84; Tim Gainey 4, 46; Goddard Automotive 95; Jeffrey Isaac Greenberg 132-133; Ken L Howard 32; imageBROKER 53; Interfoto 10, 50; Patti McConville 40; Nelson Art 112; Martin Norris Travel Photography 8; NZ Collection 88-89; Panther Media GmbH 61; Jamie Ray Images 80; Reuters 38-39; Ghigo Roli 16; ScreenProd / Photononstop 122-123; Stefano Senise 76; Sipa US 11; Bill Spengler 19; Mr Standfast 43; STphotography 58; Phil Talbot 94; UPI 149T; Vintage Archive 107

BAUER MEDIA: 116-117, 131L, 131B; Tim Keeton/Impact Images 20

GETTY IMAGES: 140-141; Chip East/Bloomberg 96; General Photographic Agency 105; Tria Giovan 23; Bob Greene/The Enthusiast Network 36-37, 74-75; Rianne Hazeleger / 500px 63; Randy Holt/The Enthusiast Network 60; Burhaan Kinu/Hindustan Times 146; Josep Lago/AFP 130; Manchester Daily Express 109; David Paul Morris/Bloomberg 128, 149B; Darryl Norenberg/The Enthusiast Network 35, 71; Popperfoto 52; RacingOne 102, 104, 106, 113, 114; Ted Soqui/Corbis 143; Edward Wong/South China Morning Post 93

HARLEY DAVIDSON MOTOR COMPANY: 153, 154-155

IAN KERR MBE: 15, 72, 111

MARY EVANS PICTURE LIBRARY: Classicstock / Photo Media 29

SHUTTERSTOCK: Julien's Auctions 56-57; Kilmer Media 48; MaggioPH 150-151; National Motor Museum 13, 17, 24-25, 78-79, 91, 98-99; Otomobil 145; Cristi Popescu 142; Pierluigi Praturlon 127; SunflowerMomma 66; Shawn Thew/EPA-EFE 138; Warner Bros/Kobal 124

WIKIMEDIA COMMONS: 134

Every effort has been made to acknowledge correctly and contact the source and/or copyright holder of each picture any unintentional errors or omissions will be corrected in future editions of this book.